INVISIBLE
to
VISIBLE

The Introverts Guide To Showing Up
Online As A First-time Entrepreneur

JULIE LAVIA

Table of Contents

Introduction

Are you a first-time introverted entrepreneur who wants to break into the online space but does not know where to start or even how to show up confidently?

As a fellow introvert, Julie Lavia found it challenging when she started working online about 4 years ago. She felt having to show up on social media was overwhelming and exhausting. She realized there are so many ways you can share your business without being on camera everywhere.

She pivoted from being a virtual assistant to a visibility strategist and coach to help other women show up more authentically and organically while doing it scared. She broke out of that shell and took gradual milestones to overcome her fear of promoting her business.

In this book, Julie writes about the strategies she has learned to help her market her business. She now wants to show other introverts leaping into the virtual world how to do it more easily.

Julie demonstrates how first-time female entrepreneurs should start with these important components such as

- Knowing their niche
- Branding
- Ideal Client
- Online presence
- Market research

This book is separated into 10 chapters that break down how to be visible as an introvert while using.

- Social media
- Camera confidence

- Faceless marketing
- Written content
- Storytelling
- PR Marketing
- Digital Products & Email Marketing

Julie has started her own membership for introverted female entrepreneurs to help them boost visibility in the online space. The Visibility Lounge is a membership mastermind to help women who struggle with being seen and need the extra hand holding to implement new strategies and to be held accountable in their daily goals. Julie believes strongly in building community and she helps her members network and collaborate by showing up to her weekly networking sessions to meet their potential client. If you want more information about how to work and connect with Julie, you can reach out to her by joining her facebook group https://www.facebook.com/groups/shemeansimpactnetworkingandcollaboration.

You can also connect with her over on

Instagram: https://www.instagram.com/julielaviacoaching to find out more about her business and membership!

CHAPTER 1

Misconceptions About Introverts

Introvert: *A shy, reticent person often keeps to themselves and dislikes being the center of attention due to their reserved and timid nature.*

Often, introverts do not get to be seen and heard because they do not know how to use their voices. Introverts are not confrontational and fade into the background to avoid conflict. However, based on my experience, not all introverts are quiet and shy. Some introverts love being in the spotlight and are the furthest from being shy. Introverts like myself can also have some extroverted traits because we enjoy socializing and meeting people. However, most of us prefer intimate settings and only spend a few hours at a time before we drain our social battery to the end.

There are many misconceptions about introverts. Here are some questions and answers that introverts face regularly.

Why are you so antisocial?

Many Introverts get asked this question all the time!

Why do you want to avoid our in-person meetups? They are so much fun! You will get to meet this person and that person, and socialize with other important people who can help you boost your business!

That is why we are known as introverts: We have a habit of being antisocial!

Don't take it personally. It's nothing against extroverts, but we prefer to be immersed in our own company. The idea of being around more than two or three people in one room can sometimes overwhelm us!

As an introvert, I wanted to have a lot of friends and would force myself to go out to be social because I thought that was what society expected us to be!

However, after years of over-exhausting my social battery to keep up with my friends and their friends, I realized I am just as comfortable staying home on a Friday or Saturday night watching Netflix and chilling in my bed while scrolling on social media.

Nevertheless, to be visible online, you must be social rather than overdo it. In other words, you have to start building connections with other people so they know who you are and how you can be of service!

There are a lot of virtual networking platforms online where all you have to do is network with each other's profiles without actually having to meet people and join activities in real life. Some of these platforms include:

Alignable: You can open a profile and connect with other communities. Fellow entrepreneurs can refer you. You can also use a private messenger to chat with people you meet. You can create a business profile to promote your services or join communities for more exposure.

LinkedIn: LinkedIn is a top-rated platform for business connections and referrals. You will often find ads for virtual events and for individuals who are coaches, business consultants, therapists, and online service providers who are looking for higher ticket 1-to-1 clients. This platform is excellent for older business solopreneurs who want to avoid being involved in a noisy social media platform but are looking for genuine business connections and referrals done by profile matching.

Why do introverts hate being in the spotlight?

In the online space, they often prefer to work behind the scenes by planning projects, mapping out ideas, and completing tasks for other business owners.

They love to show up for your business just as much as they love showing up for their own business, as long as they do not feel pressured to keep posting their faces everywhere.

Some introverts can be very camera-confident and camera-friendly. However, many introverts prefer to avoid the camera when they can. There are ways for introverts to still appear on camera without showing their face directly, which I will discuss later.

In addition, I would like to offer some tips to introverted first-time entrepreneurs on overcoming their fears of camera shyness!

Why some introverts fear showing up on camera and what to do to get started

Some introverts fear showing their true feelings and being too vulnerable to the public. We introverts are often private creatures who are only really comfortable around people we know, and that is when you see our true personalities.

It can take time for introverts to get used to showing up authentically in front of an audience with whom they have never had an actual conversation. This is why most introverts avoid creating video content in which they must stand before the camera to speak or behave out of their comfort zone.

They are too nervous because they fear being judged or not liking how they look or sound on camera.

As an introvert, I struggled to make my first live video on social media. The first time I did a live show, I was so incredibly nervous, fumbling with my words, and I couldn't look directly at myself on camera because I was silently judging my looks. I hated my facial features, the way my voice sounded on camera, and my eye contact was so poor. After the Facebook live was completed, I started crying because I felt embarrassed and disappointed with myself.

I should have felt relieved and happy that I had completed my first virtual live before joining a 1k-member Facebook group. However, deep down, I told myself, *"I don't want to do this again!"* I felt unprepared and insecure because I didn't give it my best.

If you are an introverted first-time entrepreneur ready to go live, here are some beginner tips to get you practicing to become camera-ready.

- **Transform your mindset:** Mindset is critical; you must shift your perspective to put yourself out there more. If you use platforms such as TikTok, Facebook, and Instagram, your audience needs to see more of you. Therefore, if you want to prepare beforehand, follow these next steps.

- **Create a separate Facebook group to practice your camera skills:** If you want to boost your visibility, your audience will have an easier time relating to you if you show your authentic self. Being able to show up on camera can be challenging, so creating your own Facebook group and practicing going live with few or no members to get used to seeing yourself recorded is an excellent first step.

- **The same goes for Instagram:** You can create a close friends group and practice recording your videos there, so only people you know can see them. Utilizing these platforms in this way can allow you to show up, make mistakes in front of a minimal number of people, gather feedback from those people, and use those tips and tools for self-improvement.

- **Practice your speech and eye contact before you create your official live:** Some introverts, like myself, are only sometimes comfortable with public speaking, which can cause us to mess up our recorded content and lose camera focus. Write down your bullet points or notes so your speaking can flow naturally on camera and help you get your content across to the ideal audience.

If you struggle with maintaining eye contact, try placing a sticker next to the camera lens to keep your eyes focused in one area. A public speaking coach gave me this tip. I initially struggled with eye contact because I hated being in front of the camera, but now it works well when recording a video or interview.

CHAPTER 2

The Benefits of Working Online as an Introvert

Working online can offer introverts many great benefits that align with their strengths and preferences, allowing them to remain relevant in the ever-changing online world.

Flexible working environment: Since introverts love to be alone, it is natural that they will want to work quietly at home or in a quiet cafe with minimal or no distractions. Once they are in the comfort of their environment, they can work at their own pace, take much-needed breaks to recharge, and engage only in virtual meetings.

Reduced social pressure: As I mentioned above, meetings with others can be held virtually and at different times of the day. Introverts can easily get burnt out with daily face-to-face interactions, and the ability to interact through a computer screen is golden for introverts. If we are not in virtual meetings, you can see us using other forms of communication, such as emails and messaging platforms, that don't require us to engage every minute of every day.

More autonomy: Most introverts thrive in the workplace when there is less supervision and more independence to manage their work. Some Introverts can be loners and need their breathing space; otherwise, they get easily overwhelmed and can underperform their tasks.

Without constant supervision and interruption, we can do our best to meet deadlines, stay organized, and plan accordingly at our own pace without pressure. This was one of the reasons I chose to work remotely: I prefer to be on a consistent and structured schedule.

Improved focus: As mentioned above, introverts like to work on their own schedule without facing pressured deadlines and constant supervision. Therefore, without the noise from a busy atmosphere, introverts have a higher chance of sticking to their daily schedule due to their better attention span, which leads to higher productivity.

Diverse opportunities: When working online, you are exposed to various online job opportunities from which you can choose. For instance, many introverted female entrepreneurs like myself usually start as virtual assistants.

For those who don't know what a virtual assistant is, it is a personal assistant for online business owners. As a VA, you are responsible for completing tasks that your client has assigned you. These tasks can be various or related to your job niche.

What's great about an introvert becoming a VA is that you work remotely at your own pace and schedule, and you may never need to meet with your client to submit your work. Also, VAs are required by companies worldwide and they can work anywhere in any time zone. You can work for only solopreneurs or with an agency of other VAs.

Enhanced communication: Introverts can improve their written communication skills by creating newsletters and daily content for their social media channels. Eventually, they may want to practice their verbal skills by joining more networking calls and showing up on video to engage with their audiences once they feel more at ease and have the tools in place.

Work-life balance: When working online, you have more freedom and flexibility for improved work-life balance. You can create your schedule around your family, friends, and activities. You can have as many or as few online jobs as you want, and you won't feel the physical effects because you can work at your own pace, granted that you still have to meet your deadlines from your clients.

How my introverted journey influenced my online business as a visibility coach

As you already know, I am a true introvert. However, I spend most of my days at home in virtual networking meetings lasting from 30 minutes to 1 hour to network with different business owners. It feels great to be with other like-minded women who just get it, and I'm only sometimes obliged to sign on if I have any social energy that day.

Was I always an introvert growing up?

The answer is yes! It is just that I didn't realize until I was much older that I was very socially awkward. If I was in a room or a party with a large group of people, chances are I was sitting alone, not speaking much to anyone. I was never comfortable in crowds; they needed to be one-on-one, smaller, and more intimate group settings.

You would never find me at work or in school with a bunch of friends because I was always a loner and, somehow, back in the '90s and early 2000s, we were taught that if you wanted other people to like you, you had to think and act like them, so they would be your friend or accept you into their group. The thing is, I had to learn a hard lesson because I was reticent, shy, and only interacted a little with other kids in my environment. I was bullied for being different and not fitting in with the so-called norm.

It took me a long time to be comfortable in my skin and to accept that this is who I am. I did not need to please anyone, and being an introvert can be a superpower and inspire others to embrace their individuality.

How does being an introvert play out in starting our online business?

I was tired of the regular full-time workday and felt that my job in the healthcare system was no longer serving me. I felt like I wasn't moving forward, and I had always wanted to be self-employed to live my dream life.

My dream of becoming an entrepreneur started in my early 20s, and I have never given up on trying. I fell in love with that concept, and I started learning about the value of working online.

I jumped on the idea right away, and I started to learn about becoming a virtual assistant. I wanted to avoid facing people and daily traffic to make a dollar. I was tired of feeling and playing small, not being appreciated for my work, and feeling like a number.

I saw how these women entrepreneurs were crushing it by working on their laptops in a foreign country, sitting by the pool, and drinking a cocktail. I admit to having been low-key jealous of them living this lifestyle and working only 3–4 hours per day while I was making someone else richer and my health was declining.

Once I started as a virtual assistant specializing in content creation, I realized it was not for me. I pivoted to coaching, and I felt more in place. I had a knack for teaching others, especially since I wrote blogs and provided educational content; it was a no-brainer.

I learned about the industry and where I could be of the most help. I improved at showing up live to engage with my audiences on social media. However, I realized for first-time entrepreneurs this might be difficult. I wondered where and how they could fit in this space amidst the oversaturated content. If I wanted to be seen, I had to get on there and do what everyone else did.

I realized later that, as introverts, it can be challenging and scary to put ourselves out there, and the worst part is not even knowing where to start. This gave me the idea of becoming a visibility coach to first-time introverted entrepreneurs! Now, I help women in my Facebook community stay visible through networking, community, and collaboration. I provide offers to help them streamline their business, so they can always show up through automated systems.

CHAPTER 3

Things You Should Know as a First-Time Entrepreneur Before You Launch!

It can often be overwhelming to figure out where to start an online business. This is especially true if you are an introvert struggling to find your place to shine in this busy space.

However, before worrying about visibility, you must know some crucial steps to help you better navigate your online presence. Here are some tips to help get you started in the right direction.

- **Niche:** You must know your niche. You cannot serve everyone, and everyone will not need your message. If you want to narrow down your niche, ask yourself these questions before you jumpstart your online career.

  1. **What are you passionate about?** Find out what lights you up and what you could talk about for hours even if you weren't getting paid to do so.

  2. **What questions do you get asked the most?** What do people love to receive your help with? Find something in which you have the most knowledge and expertise, and use this to hone your niche.

- **Branding:** Your brand will represent your business, so you need to practice authenticity with your brand. Since you are a first-timer, you should specify that in your brand messaging when introducing yourself. Focus on working on your true self and how you would like to show up for your audience based on your comfort level. Your branding influences your online presence, and as an introvert, you are most likely comfortable showing up on professional websites, blog posts, and faceless videos.

- **Perfect customer/Ideal client avatar:** As I mentioned, you cannot serve the entire universe. This is why you need to nail your niche and find out who your ideal customer/client is. For example, to help other introverts, you must first know their perfect profile: gender, age, occupation, and income status. This will determine the exact target market that you are trying to reach. To know how to help your ideal customer, you must know what their pain points are. *Pain points are what your audience is struggling with.* Ask yourself:

  1. *What are they trying to solve?* What is the recurring problem that they cannot find the answers to? How can you help them solve that problem?

  2. *What is the missing gap that your audience keeps running into?* What are your competitors saying, and what are they not saying? Whatever they are not saying, you can consider that to solve your audience's problem.

- **Online presence:** Showing up on social media platforms is one of the most critical areas of our business because this is how we find leads, network, and make potential sales. Since I'm an introvert, I started with an introvert-friendly site like Facebook. Since you do not need to show your face to connect with others, here are some ways I used Facebook to leverage my business.

  1. *Join like-minded Facebook groups:* Once you have a niche, a professional website, and personal messaging, you can search FB groups where your ideal clients hang out. You should start engaging with posts to build relationships and provide value in your niche to build trust.

  2. *Create your own Facebook Community:* I created my community and invited other business owners to join.

Doing so helps me nurture my audience, and I can choose to go live or have one-on-one interactions with select members who need more of my help!

3. *Join or create small networking groups:* Introverts thrive in smaller groups. I created virtual networking groups within my FB group to improve my camera confidence and have deeper conversations. If you don't plan to host your networking events, you can join other niche-oriented meetups where you can provide value and promote to your tribe, which simply gets you and your business.

4. *Create a website:* If you want more credibility and visibility, you should create a website to showcase your brand. Your website is your virtual business card that stores all the information about you and your business. You should have your website set up with your ***About page, Services, Portfolio, Resources, and Contact Us.***

Overall, choose the platform that makes the most sense for you. You can try out a couple of platforms to see which ones you prefer, but I think it is ideal to master at least one platform before you appear on multiple.

- **Market research**: When you want to find out who your ideal client is, you must do some market research. Speaking to a large group of people may feel overwhelming as an introvert, so if you decide to use Facebook as your primary traffic source, here are some tips on how an introvert can conduct market research!

    1. *When joining another Facebook group,* if you see other people in your niche talk about a pain point related to what you do, you can simply send them a DM or a comment under their thread to provide value. This strategy could help build a relationship with this person and lead to a more

extensive conversation where you can talk about helping them. This can work for introverts because they can engage in one-on-one conversation and feel free to accommodate only one person.

2. *If you have your own Facebook group,* you can create polls to ask your fellow members what they would like to learn more about or what they are struggling with. This can work well because the members and followers are already part of your world and trust you and your content. Introverts must be comfortable in their environment; it can take a while before they feel at ease around strangers. This is why it is ideal for introverts to have their community to build and develop a closer relationship with their potential clients.

How can an introvert who is a first-time entrepreneur find members to join their community?

This can be scary at first, as so many people online are trying to grab one another's attention. However, don't let this overwhelm you; keep looking for your people. There is enough space for everyone to shine!

If you want to start by building your community, you can ask people you already know, such as friends or other people you have made past connections with, to join. This can give you an excellent start to getting some members inside.

You can also ask those members to refer friends or other business owners they know who they feel would be an excellent fit for the community. This can be helpful if you are too shy and reserved to engage with people you don't know. ***You can give an incentive by telling your members that whoever refers the most people will win a giveaway or raffle.*** This can be a free service or a virtual gift card, as those are common choices to provide to members.

Once you are comfortable with having many members in your Facebook group, you can start engaging with them by hosting networking events to get to know them better.

Why networking events are beneficial to networking as an introvert

Virtual networking: Some introverts avoid face-to-face meetings, especially within a huge crowd. Virtual Zoom events can host small groups, making it easier to build better rapport. Once that rapport is built, you can easily ask for market research and create services based on various wants and needs.

Work on your public speaking and camera confidence skills: Networking events are perfect for improving communication if you fear speaking on camera or within a group.

It is your choice to keep your camera off during the event, but you should get used to keeping it on for better connections. By keeping your camera on, you can encourage others in the room to turn on their camera and answer questions to boost engagement.

If hosting an event alone feels overwhelming, you can always get a co-host. The co-host can share hosting responsibilities with you, and if the co-host is more comfortable with social interaction, they can have more authority in managing the group. Start using a co-host to bounce ideas, help you provide value, and keep members interested.

CHAPTER 4

Visibility and Social Media as an Introvert

What exactly is visibility?

According to the dictionary, *visibility is the state of being able to see or be seen.* When you show up, you can impact and influence others.

One of the most effective ways to stay visible as a first-time entrepreneur is to leverage social media platforms.

Why social media?

While some individuals prefer to avoid existing in the world of social media, if you plan to work online, chances are you will need to register for a platform so you can start building a brand.

How can social media be effective for your online business?

Social media can play a huge role in whether you can successfully grow your business. While it is not the only effective tool available to gain visibility, social media alone has helped millions of online business owners make six-to-seven figures.

Here are some reasons social media can be effective for your online business:

- **Communication:** You can be one click away from your next potential client or business partner. Social media makes it easy to communicate with others' personal or business profiles through voice or text messaging.

- **Collaboration:** Social media allows you to find collaboration through personalized Facebook groups geared toward your search. If you cannot find what you want, add a keyword to the search engine or form your own community.

- **Inclusivity:** You can join groups and create your own to invite people who just get it. Once you access the right community, you can start forming relationships, so they trust you and your brand. Once that trust is established, you can better sell your offers.

- **Sense of belonging:** You can find your group of people with whom you can build an even tighter community. For example, suppose you have a Facebook group or belong to other private groups. In that case, you can invite your ideal clients to private events such as webinars, networking events, and messenger or telegram chats. The fantastic thing about building private communities is creating a group based on race, gender, hobbies, goals, location, etc. You will have a higher-than-likely chance of having high engagement and the ability to convert leads into sales if that is your desired result.

How to use social media as an introvert

Social media is a boisterous place; you must make some noise to stand out. You will usually see different people showing up in other ways.

Let's discuss a bit about the different platforms that make up the world of social media and their actual uses. This is mainly for anyone living under a rock who uses few to no social platforms and wants to understand the pros and cons before getting involved.

Facebook & Instagram: Facebook and Instagram are great spaces to build a business profile that showcases your name, website, and a short bio or description of yourself, who you serve, and what kind of business you do. You can add your website and extra social links for anyone who wants to connect further with you. You can use the tools that Facebook and Instagram offer for anyone who wants to get to know you better for the like and trust factor.

Stories: The Stories feature is suitable for short-form video content lasting up to 30 seconds. You can add photos and Canva graphics to showcase business milestones, memorable wins, first clients, or sales. Stories last 24 hours before they are taken down and stored in your Facebook memories. If they are active and consistent, the majority of your audience will see your stories. The more stories you create and post, the more engagement you will have as the content gets shown to more people on your friends list.

Reels: Reels are just like stories, except they are short-form video content that plays for 60 seconds. Unlike stories, reels can be shown to multiple people who are not on your friends list since you can use hashtags to reach more people interested in the same niche.

You can use reels to promote your business by providing a hook to draw the viewer in and a call to action so your ideal viewer can click to take action.

Here are reel and story ideas that you can use to boost visibility without having to be on camera:

- ***Behind the scenes:*** You can create a video of what you are working on behind the scenes. If you are working on a course or a launch, preparing for a client, recording a podcast, video, content, etc., you can give your followers a glimpse into your daily routine. Filming BTS content shows that you can be trusted as someone who can help clients solve their problems.

- ***Tell a story:*** Talk about how you started on a path to success vs. how it's going now. You can allow yourself to be vulnerable by sharing the struggles you had when starting and how you have progressed by making leads and sales or by sharing a small win or goal. You also do not need to show your face; you can just use stock images and text overlay to create your captions.

- *Engagement value posts:* These are the types of posts you can use to keep up engagement with your community. You can ask questions to do your market research as well as create polls to get people to vote for their favorite choice. You can make your reels and stories to post motivational quotes, and affirmations, or provide a valuable tip in your niche while using a call to action such as "like or drop an emoji if you agree."

- *Market a product or service:* If you want more engagement and reach, you can create a piece of short-form video content to market your services for your business. Again, it is your choice as an introvert how much you want to show your face on camera. We will get into more details about faceless marketing a bit later. You can use a promo graphic to market your podcast episode, website, and latest offer or services to let your audience know how they can work with you.

- *Repurpose your content:* You can use different content that you posted on other platforms and repurpose the content into a reel, story, or social media graphic and caption. This is the best way to show up in different places without burning yourself out by physically going everywhere. Content repurposing is when you extract the main points from written or audio content and reuse it to create a snippet of video. For example, if you are a host of a podcast or you were a guest, you can use snippets of audio to create an audiogram to play in your reels or stories so you can promote the upcoming episode of the podcast.

- **Record daily life activities:** Your ideal audience loves to see that you are legit and authentic. They love to see your real life behind the scenes when you are not talking about your business. You can create a short clip of you at your favorite restaurant, in your hometown, with your family, and anything else that is part of your daily routine.

- *Canva graphics*: You can use the graphic design tool Canva to create a static or a carousel post. You can make these graphics to post on your newsfeed and then share them with your stories and reels as repurposed content. It has been proven that your followers will watch your stories and reel content more than your feed. Nowadays, attention span is concise, making it harder for them to sit through 60 seconds of content. You should keep your video content under 10 seconds, especially regarding reels.

Facebook and Instagram live: As an introvert, it is often challenging to show up live because we don't always feel comfortable showing our faces. If you have never gone live on your social media, here are some good reasons why this will help you boost visibility in your business:

- **Brand authority:** You are showing up authentically to show your audience that you are the real deal and know what you are talking about. People are looking for an answer to their struggles. If you have the solution to what they are looking for, people will be looking to you as their go-to expert and seeking support around their issues. If you go live on social media to provide valuable tips and tools and show your followers ways that you can help them achieve desired results, they will develop a relationship with you and your brand.

- **Collaboration:** I know it can be super scary to do a recorded live on your Facebook profile or even in your own Facebook group. Therefore, I recommend you start by doing an Instagram collaboration where you and another guest can do an interview-style video on one another's business profile. These videos will help boost viewership on both pages since all followers will get notified.

CHAPTER 5

Visibility and Camera Confidence

In the first chapter, I wrote about overcoming camera shyness to improve visibility. As an introvert, you do not necessarily need to show up on camera daily, but if you want to establish better relationships, you will like to allow people to know who you are.

I say this because there are so many people in the virtual space who pretend to be people they are not. Creating a fictitious profile without a photo is easy, so you can hide behind the screen.

If you are trying to grow, scale, and launch a business, people will want to buy from someone with a legitimate business. Therefore, it is a great idea to let your potential customers know who the person behind the brand is.

What is the worst that can happen if you never show your face when marketing your business?

The worst that could happen is:

Slow growth in building business relationships: When you are too shy to show your face, you may have a more challenging time reaching more people personally. When I was starting, I did not like to be on any social platform where I was required to show my face. Little did I know I did not have to show my face daily, but I would still need to balance the two.

I started my Instagram account with just a picture of myself, but all my content was promo-branded graphics. I was still new to the online space, and I had yet to hear the lesson about keeping up with followers on Instagram, which had more visibility.

If you don't create reels and stories and don't go live to boost visibility with your audience, chances are they will not see your marketing content, especially if you are making only graphic posts. The algorithm will only show your content to a small number of people, usually just your followers, but if you want to attract more people to your business, you will need to utilize more of the platform's tools! This was something that I had to learn the hard way because I would only choose to stick to the main feed and stories with no converting clients for my Virtual Assistant business at the time.

Lack of sales: This brings me to what I had mentioned about not getting any lead conversions. Suppose you just had a website created without photos of yourself and a social media account with no engagement activity. In that case, it could be more accessible for people to find your services if you use word of mouth or referrals.

While it is still possible to get leads and sales, you will have to find other alternatives to get exposure that do not require you to show your face. This can be through:

- Podcasting
- Blogging
- Sales Funnel
- Email Marketing
- Virtual Events

I will explore these topics more in-depth later, but I wanted to discuss how to show up confidently on camera before you are comfortable repurposing content into video format.

Camera confidence for introverts

Camera confidence can be a challenge for introverts, so I wanted to share some valuable tips, practical steps, and mindset shifts to help you feel more natural in front of the camera.

1. **Create a comfortable environment:** You should find a familiar space to reduce the anxiety of appearing on camera for the first time. When I was going on camera for the first time, I would feel more at ease recording in my bedroom. In the bedroom, I can be myself, shut my door, sit on my bed, and adjust the angles to look presentable on camera. Whether you want to hang out at the park or in your living room, you want to ensure a quiet environment with minimal distractions.

2. **Create good lighting and setup:** Ensure your lighting is bright enough and your camera setup is stable. Here are some tips on tools you need to have the perfect tech setup

   - *Ringlight:* A desktop or a standing ring light can provide extra lighting, especially in a dark room.

   - *Stable internet connection:* Ensure you are in a place where the internet connection does not lag and remains stable while recording your video. The worst case is if that were to happen, you could just edit the video for any mistakes or disconnections.

   - *Microphone:* Attaching a microphone to your device is an excellent option as it enhances the audio while speaking. Use a microphone for better sound quality if you use a laptop or a desktop. If you record using your cellphone, the sound usually has excellent acoustics.

   - *Earpieces:* When I am about to record, I like to wear earphones to eliminate background noises. While it is recommended that you find a quiet place to record, you cannot always predict when or where noise will occur. You should have some ear devices handy, such as headphones or air pods, especially if you are being interviewed by someone else.

3. **Do thorough preparation:** It is a great idea to start by writing down the key points you want to discuss on camera. Coming on camera prepared reduces the chance of anxiety and having to think on the spot.

- *Rehearse where you feel the most comfortable:* Before you go live, you will want to get used to practicing your speaking. You can practice in front of the mirror to start making eye contact.

- *Start recording frequently:* To familiarize yourself with the camera, you can create short-form clips and videos. Once you feel more confident, you can start increasing the length of your video. The more you practice, the more comfortable you feel, adding more content.

Ask for constructive feedback and criticism

There are two ways to get feedback.

Ask a friend for feedback: If you want to hear someone else's opinion before you share it, you can ask a trusted friend to watch your video. You can then start sharing it with friends and asking for improvements until you feel comfortable sharing your videos in online communities for even more support.

Be your own worst critic: If you are not comfortable showing your work yet, you can watch the video replay to catch your mistakes and learn how to improve.

Mindset shifts and visualization

You may be expected to feel nervous and anxious and have negative thoughts because of your fear of not knowing. You must help your mind get back on track to focus on the task. Here are a few tips:

Deep breathing: Before you put your camera on to record, you will want to do some deep breathing exercises to work out any nerves you may feel. You can deep breathe by doing meditations while using positive affirmations.

Deep breathing can be done by lying down on a mat on the floor or on the bed while listening to your favorite video or podcast. You can be calm, relax your mind, and remind yourself that you can overcome your fear of appearing on camera.

Positive visualization: While you are in a positive mental state and repeating those positive affirmations, you can start to visualize a virtual room full of people or just the perfect person you are speaking to.

This can give you a sense of being welcomed and feeling like your voice matters to someone. Even if you are on camera when no one is watching, you can still show up like a million people are in the room.

Join like-minded communities: You can find an online or in-person community where other introverts hang out. If you are looking for a community where you can improve your camera skills, you can join workshops and challenges that offer this to help you get started.

Accountability Partner: If you join communities, workshops, and challenges, you can ask for an accountability partner to hold each other accountable. As introverts, we love to work alone, but having another person to support us can make all the difference in our work performance.

Develop a routine:

Develop a pre-recorded ritual: Before you decide to record, aside from deep breathing and affirmations, find a routine to put you in a good mood. If you like to dance to music, add it to your pre-recording routine so you can bring positive energy. This way, you will feel alive and motivated to speak on camera regardless of whether people will be watching.

Maintain a consistent schedule: It is best that you want to build a habit when it comes to pre-recording. The more you practice recording daily, the better you will get at filming clear, concise, and good-quality videos.

Introverts often worry about showing up because we become too critical of how we look or sound. I am a great example of this; before I started my business, I never wanted to show my face on camera. I didn't like how my voice sounded or my face looked, even with makeup. However, I had my why for wanting to show up, and that was because I wanted to be a business owner to help people thrive in their businesses.

I also learned another lesson as I started getting more camera-confident due to consistency. When you show up on camera, it's not about how you look; it is about the messaging, the value, and the engagement you provide, so more people can build a stronger rapport with you and your brand.

Start sharing with others in the community: You have been preparing behind the scenes for so long that you should feel ready to post your video content for the public to view.

Find a community where your ideal clients hang out. Then, start creating video content based on what the audience wants to hear so you can solve their problems.

Embrace your imperfections: You want to have fun and not worry about whether you stuttered or are having a bad hair day. Believe it or not, viewers will find you more relatable when you show up on camera authentically. You do not need to be perfect and polished because we are imperfect. We must learn to embrace our faults and roll with the punches.

When you implement these strategies, you, as an introvert, will build more confidence and feel more comfortable in front of the camera. The best way to approach this is to take gradual baby steps and celebrate each small win.

What happens if you want to avoid always showing up on camera to market your business?

Is there such a thing as creating a business brand as an introvert without having to show their face on social media?

The answer is yes!

Many people are using a new and popular strategy called **Faceless Marketing** to convert leads into sales! We will discuss this marketing strategy in the next chapter!

CHAPTER 6

Faceless Marketing

What is Faceless Marketing?

Faceless marketing is a strategy being increasingly implemented by business owners to promote their products and services without showing their faces.

Basically, instead of showing the face and personality associated with the brand/company, a brand can be highlighted mainly by its selling product and customer service experience.

Although many individuals in the online space use Faceless Marketing extensively to build their brands, this strategy is especially useful for introverts.

Here are some benefits why Faceless Marketing can benefit introverted entrepreneur

Less personal exposure: Introverted entrepreneurs can easily build their brand without having to put themselves at the forefront of their company. This is ideal for an introvert who wants to minimize the need for more personal visibility and making public appearances. Another type of Faceless Marketing that can also be effective is having a website.

A website is like a personal business card, where you can add your logo, business name, brand fonts, and brand colors, along with what your website copy says about you and your business. If you have your website optimized and promoted on the right platforms, your systems can automate the sales process for you to generate leads and sales.

Challenges: While building a Faceless Marketing brand has some benefits, one challenge can be needing to work harder to stand out in a crowded social media space.

To maintain high engagement and stand out in the noisy space, you must be more creative with content strategies since your audience cannot form an impression of the individual's personality.

Quality content: When using Faceless Marketing, introverted entrepreneurs can focus on creating compelling, informative, and valuable content. This helps them showcase their expertise and resonate with their audience.

Here are some tips on how to get started with creating Faceless Marketing content

1. **Understanding and defining your objective:** First, you will want to ask yourself these questions on what you want to achieve with your faceless marketing:

   - Are you selling a product or service?
   - Are you looking to boost your lead generation?
   - Do you want to bring more visibility to your brand?
   - Do you want to focus on customer engagement?

2. **Researching *your audience needs*:**

   - What are their interests?
   - What are their pain points?
   - How can you help them solve their issues?

   Once you conduct market research, you will have a better idea of what kind of content to create for them that will spark high engagement. You can even look at your competitors to see what kind of content they are using to convert into sales. For example, you can check out other faceless brands and the current trends they are using to market their business.

3. ***Develop your brand voice*:** Ensure that your brand voice is strong and consistent across platforms. Decide if you want your

tone of voice to come off as humorous, quirky, serious, or professional.

4. ***Use of visual elements:*** You will want to create visually appealing graphics based on your brand visuals so your future customers can identify your brand and content. Here are content ideas you can use that are faceless and can help you clarify your brand messaging.

   - **Graphics and illustrations***:* Use custom graphics with a graphic design tool. The best tip is to create a ready-made template that you can reuse with your brand colors and fonts already saved, so you don't need to spend hours recreating graphics.

   - **Animation and motion graphics:** Use animated videos and motion graphics to create higher engagement with your viewers.

   - ***Stock images and videos:*** high-resolution stock images you can find on royalty-free stock photo sites where there are no faces involved. There are websites that provide free and paid stock images that you can use as graphics to display on your platforms, such as:

     o Pexels
     o Unsplash
     o Istockphoto
     o Pixabay
     o Freepik
     o Picmonkey
     o Canva

There are different kinds of Faceless content you can create for more visibility in your business.

Polls, quizzes, and surveys

If you are looking to conduct market research from your audience, the best way to collect data is to create a poll on Facebook and Instagram.

If you want to create a poll on Facebook or Instagram, you can go to the post section and click on Create a poll in the drop-down menu of options. This has always been a helpful tool whenever I want to ask a question to my audience for feedback. They can take a vote, and whichever questions resonate with them the most will be the one the author uses to solve the audience's problem.

Quizzes and surveys are another productive tool for gathering market research, feedback on your services and products, or information for an upcoming event or launch.

Offering a contest or giveaway is the best way to get people to participate in a quiz or survey. For example, you can tell your audience that if they fill out the survey or quiz, they will be automatically entered to win a gift card, gift basket, or a free service that you normally would charge for. Providing a contest or giveaway is a great incentive to get people motivated because there is a reward involved.

Educational and informative posts

Educational posts are the types of posts that provide educational value to teach your audience key takeaways that can change their lives. For instance, if you are in the health and fitness field, your content should be geared towards healthy living. You will create a post with either an animated or stable image that will have a hook to draw in viewers, along with a text that reads something like this:

Hook: *Here are 4 ways to improve your gut health without having to diet*
First Slide: **Drink more water daily**

Second Slide: **Avoid sugary foods**
Third Slide: **Eat smaller portioned meals**
Fourth Slide: **Eat moderately while snacking**

Call to action: 'If you want more tips, follow me on Instagram for more!'

Now, this is not exactly the proper guide to gut health, but it is intended to provide an example of how to provide educational content.

You are not only providing value that they can implement, but you are also showing them that you are the go-to expert in their niche and can solve their issues with gut health.

This type of content would be best presented on a Reel, Stories, or Static post on Facebook and Instagram.

There are also a few other posts you can use as well to get the educational and informative content across, such as:

Infographics: A one-sheet post that shares informative content with text and graphics to highlight tips and provide educational value.

Webinars & live workshops: You can create webinars on educational topics surrounding your product and services to give viewers more insight into how investing can transform their lives. You can offer a question-and-answer period and serve a multitude of people without choosing to show up on camera for the entire webinar.

Live workshops: You can create private free or paying workshops where they are more intimate and focus on teaching a specific topic in your niche. This goes back to the example of the gut health post from earlier, you can use the workshop to elaborate more on the content if they need more information. You can host on Zoom or on your platforms, such as Facebook, Instagram, and TikTok.

Product centered content

If you are promoting products to sell, you would want to make that product the main attraction of that post and showcase the benefits of why this product should be considered as an option in your life.

For instance, if you are selling meal prep services, you can list the benefits, such as:

- If you are a busy mom, you can save time by preparing your meals in just under 30 minutes. No more slaving over a hot stove and missing out on spending time with your kids!

- You can choose from a variety of different recipes so you do not have to eat the same boring meals night after night.

- You can order up to 3 meals per week to save money on shipping. In addition, you can choose to skip a week's worth of meals if you do not want to purchase them that week.

The features of meal prep would be:

- Each meal comes with a recipe card, which features images and prewritten instructions to take the guesswork out of cooking, especially if you are a novice in the kitchen.

- Each menu includes all of the ingredients, from pre-packaged spices to pre-chopped vegetables.

- All packed boxes are delivered at a cool temperature to preserve cold meats.

You would want to use high-quality video images and videos to display on your platforms!

How-to guides: You can use the same meal prep product and turn it into a video tutorial showing you or the customer preparing the meal from start to finish.

You can also repurpose that video tutorial into a step-by-step guide or ebook for those who prefer to read content instead. The guide can include images taken from the original video to showcase how to use the products effectively.

User-generated content

Customer reviews and testimonials: If you are a first-time entrepreneur and want to gain more credibility when promoting your business online, it is best that you provide social proof or customer reviews from past clients.

You can start showing up by offering your services for free or as an intern in exchange for a review or testimonial. This will help you build up your credibility as a legitimate business owner much quicker, and those you worked for could potentially become paying customers or refer you to others down the road.

Takeaways from using faceless marketing for your business

If you want to be successful at Faceless Marketing, you will want to focus on quality content. Since you are not showing your face, you will depend on your brand messaging and brand consistency to attract the right people to you. Ensure that you are leveraging your visual and written content to enhance your engagement. The challenge can be the idea of building a personal connection since it is faceless marketing.

Faceless Marketing can align well with the strengths of introverts because introverts are able to grow their presence without the stress of personal exposure.

However, it can also be challenging to build a personal connection because they are not showing their faces. If they do not show their faces, it is unlikely that a customer will want to work with them because they could assume they are in line to get scammed.

This is why I suggest that if you want to dabble as a first-time introverted entrepreneur with Faceless Marketing, you should at least include some professional photos of yourself and add some overlaying text to talk about your business.

You can use a strategy where you post one picture of you per week so people can at least know that they are dealing with a real person behind the screen. I would say when you are gradually more comfortable, you can start making short-form videos once you gain more camera confidence.

Now that we have discussed Faceless Marketing and visual content on social media, I would like to explore the power of written content for extra visibility.

CHAPTER 7

Written Content for Visibility

Why is written content beneficial in the online space?

Written content is a valuable tool when you are just showing up online for the first time. Here are some reasons as to why it is crucial as a business owner:

- **Engagement:** Engaging content allows visitors to stay on the site, explore other relevant content, and possibly return in the future. When you have written content, you are creating a gateway for positive interaction with shares, comments, likes, and discussions.

- **Authority and trust:** When readers find valuable and educational content, they are more likely to trust the source and want to keep reading its content.

- **Content marketing:** Written content is the catalyst for content marketing strategies. If you want content to convert leads to sales, you should try blogging, newsletters, social media content, and storytelling. We will go into more detail about each one in a bit.

- **Brand voice and identity:** Through style, tone, and language, brands can have separate identities to attract and connect with their audience on a deeper level.

What are the different types of written content for introverted entrepreneurs to post in their businesses?

Blogging

Blogging is an informative article written to educate readers on a product or service. Your blog posts are made up of the following written content:

- **Product reviews**
- **Affiliate marketing**
- **FAQ on a product**
- **How-to guides**
- **Listicle**

Blogs are great to host on your website because they can give your brand more visibility and more insight into your brand. The purpose of a blog post is to:

- **Boost SEO:** Search engine optimization (SEO) enhances visibility by using specified keywords that rank on Google. In other words, if a reader is searching for a specific keyword, such as '*gut health*,' Google will pull up articles and other blog posts that contain those keywords.

- **Brand authority:** Blog posts can help you build brand authority by allowing you to write about your knowledge and expertise in your niche. When you write a post explaining why gut health is important to healthier living, you are listing out the main points to help your audience get answers to their issues with gut health.

Why blog posts are great for introverted entrepreneurs

Creative outlet: Blog writing allows introverts to be creative in expressing their interests, passions, and personal perspectives to a broader audience.

Happy medium: Introverts are in a happy place when they can write to express their thoughts and ideas without feeling the pressure to interact in real-time.

Influence and impact: When an introvert writes a well-written blog post, it can inspire and influence others by impacting their businesses. An introvert can take the time to formulate proper comments instead of having to be put on the spot.

Career opportunities: Writing successful blogs can open up opportunities for freelancing jobs, consulting, speaking engagements, and book deals.

Flexible schedule: Blog writing allows introverts to work steadily at their own pace. The flexibility allows them to write when inspiration and productivity take over.

How to start a blog post as a first-time entrepreneur

- Find your niche
- Conduct market research
- Decide on a blog name and website domain

Once you have these 3 items in place, you will want to learn how to write a proper blog post to attract your ideal reader. Remember, the main goal is to blog about your expertise so the reader can take the next actionable steps to learn how to get more value from you.

- Add a catchy headline with your keyword as the first word in your title.
  Example: **Gut Health, 5 Myths debunked, and 5 ways to a healthy gut.**

- Add your keywords like *gut health* at least once in the introduction paragraph so Google can recognize what the blog post is all about. This is what makes the post SEO-friendly.

- Separate your paragraphs with headlines to break up the structure of the blog post. This would be ideal on a listicle-style post, such as the example post from above. You will list the 5 myths with a different headline and a paragraph under each one explaining the debunked myth in detail. Followed by the proper 1 or 2 ways to a healthier gut in a short paragraph.

- Add a conclusion to wrap up the post, providing a brief summary and a call to action such as: *To learn more about Gut Health, download this free guide here.*

Newsletters

A newsletter is an article used to engage email subscribers in an email marketing campaign. If you are a complete newbie, you are probably confused by what I just said, so I will break it down a little further for you.

When visitors come to your website, they will browse through your resources, such as a free guide, ebook, and templates. If they want to download it, they will have to sign up through your opt-in form.

The opt-in form captures their name and email address and adds them to the host's automated email marketing platform in exchange for the free item. The email marketing platform then sends out an automated welcome sequence to welcome them to the community and nurture a business relationship with you.

The purpose of reading a sequence of newsletters is to develop a know-like-trust factor with you so that you can be upgraded to their higher ticket offer.

How do you create newsletters?

Newsletters have to follow a nurturing sequence to take the customer through a journey. You start with them as warm leads since they are already on your list, and you want them to convert to hot leads eventually.

It starts with an automation sequence. New email subscribers will receive 6–8 emails every other day until the sequence is completed. Once it is finished, they will be added to the regular mailing list, where the frequency of emails per week is usually much less.

The regular email list is just to establish a relationship with the subscribers and keep them up to date on the latest news, offers, and changes in their business. Some subscribers may not buy any of your offers, but the right one can make them want to purchase.

Therefore, you need to check your analytics of your open and click-through rates before you decide to clean off inactive subscribers on your list.

Introverts can benefit from creating newsletters because it is similar to blogging. By asking engagement questions in your newsletters, you can create one-on-one relationships with your subscribers. You are not obligated to show your face because email marketing is considered almost like Faceless Marketing.

- You can create a relationship with your subscribers with the written content and they can get to know you through the art of storytelling through email sequences.

- If you want them to stay further in touch with you, you can direct them with a link to follow you more on social media for more tips. You can continue to build the relationship there as well through post comments or DMs.

Social media content

Showing up on social media can be challenging, especially as an introvert, because there is so much noise, and you can get overwhelmed by where to start. However, you need to show up in a way where you are feeling comfortable. For introverts, you can start by creating carousel posts, which are slide decks of different written content explaining '*the 5 ways to get over belly fat.*'

For example:

The first slide will have the topic headline such as the **5 ways to get over belly fat.**

The second slide will have the 1st headline such as **intermittent fasting** and the reasons why it works for belly fat.

The third, fourth, and fifth slides will discuss other ways to get over belly fat.

Some things you should know when creating content

Keep your graphics and images on brand: Ensure that you keep the same brand colors and fonts so that your audience is familiar with your brand. This way, they can see your posts on their feed on Instagram or Facebook and recognize your brand visuals without looking at the creator.

Add some relevant images to your posts: If you are talking about getting rid of belly fat, it would be great to add some visuals to back up the content. Intermittent fasting pictures would be a great example, such as a before-and-after or a chart to follow.

Captions: Captions are written content that accompanies a photo or video on social media. They can be used to add additional content to video or written posts, tell a personal story, or add humor to the posts.

For example: You can add a photo of reducing belly fat with images of you doing a workout or showing you cooking a healthy meal. The caption can read something like this:

Are you looking to get rid of belly fat before the summer starts?

Personal story: I used to be ashamed of walking around on the beach because I did not want to show my belly through my one-piece bathing suit!

I would look around at other women wearing their bikinis in confidence, not having to worry about their bodies being judged and criticized for looking out of shape.

I got fed up and decided that this summer, I would do what I could to work on better gut health.

No more mindless eating or lack of exercise as an excuse for poor gut health!

If you want to get rid of that annoying belly fat so you can rock that skin-tight bikini at your next beach holiday, here are some tips to get over belly fat:

1. *Intermittent fasting*
2. *Swap junk food for healthier snacks*
3. *Exercise regularly*
4. *Keep a journal of meal prep*
5. *Stay hydrated*

This is not a list for getting rid of belly fat but a mere example of what social media captions can look like. If people prefer to gain value through watching a video or reading written content, there is something there for everyone.

AI & Chat GPT: If you have trouble creating written or visual content, you can always rely on sophisticated AI tools

Chat GPT: An AI-generated site where you can ask questions, search for ideas for social media content, headlines for blog post topics, and how to formulate newsletters and sales page copy to create an offer to sell.

Chat GPT is a time saver when it comes to creating content. However, its pitfall is that AI-generated tools can make it challenging for other service providers to get hired to create content due to AI's popularity.

If you are going to use AI-generated sites, you would need to add your own content or paraphrase the content to avoid copyright infringement.

CHAPTER 8

Storytelling

Storytelling is crucial when you are in the online space. Here are reasons as to why storytelling can benefit your business.

- **Engagement:** Telling a personal story tends to capture your audience's attention. People can resonate more with your story because it helps stimulate emotions and imagination. Your storytelling makes content more memorable and engaging with your followers so that they will want to come back to learn about you.

- **Connections:** Personal interaction can be limited in the online space, therefore, creating personal stories helps bridge the gap by making interactions feel more personal and human.

- **Trust:** It can be challenging to believe which brand or business is authentic and legit. When brands and individuals share genuine experiences, it creates credibility and transparency.

Here are some tips to get you started with storytelling

1. **Understand your audience:** You must be able to relate to your audience. Consider their demographics, preferences, and the types of stories that resonate with them.

2. **Define your purpose:** What is your story trying to achieve? Do you want to inform your audience? What about entertaining them with a funny story? Do you want to persuade them with your latest offer? How about inspiring them to do better or stay motivated? You want to clarify the main message that you want your audience to remember.

3. **Finding your story:** If you plan to use more storytelling, you will want to revisit personal experiences, how you started your brand, and social proof from past customers, or collaborators.

4. **Clear structure:** Follow a clear structure such as an introduction, conflict and challenge, and resolution or conclusion.

5. **Be authentic:** Did you know that authenticity builds trust? Your authentic stories should reflect real-life experiences, values, and voice.

6. **Use of multimedia elements:** You would want to add visuals to enhance your story for extra credibility. You can use videos, audio clips, and visuals to show examples and to back up important points in your storytelling journey.

7. **Engage and interact:** You want to encourage active participation and interaction from the audience. If you are doing an Instagram live or a Zoom presentation, you should ask questions, invite your audience to make comments in the chat, and allow your audience to share their own stories.

8. **Analyze and adapt:** You would want to adapt your storytelling strategies based on how your audience reacts to the amount of engagement and personal feedback.

How to apply these steps in your business

For example, if you want to explain your journey in starting and building your own business as an introvert through a webinar.

- **Audience:** Your target audience is young introverted women who want to start their online business in digital marketing.

- **Defining the purpose:** To create excitement and get potential leads for the launch of your new hybrid coaching program.

- **Finding out your story:** This is when you will talk about who you are, what you do, who you help, how you started your business, and what challenges you had to overcome in starting your coaching program.

- **Clear structure:**

  Beginning: Get them excited by asking engaging questions such as where they are located in the world and questions that require them to respond with a yes or no or a particular number. By starting to engage your audience in the beginning, there is a higher chance they will want to stay to keep engaging with you.

  Middle: You will talk about where you went wrong in your journey or what the problem in the industry is. You can explain that when you started your business, there was a huge gap in the coaching industry for introverts starting to learn the steps of digital marketing. If you invested in the wrong courses and coaching programs, you can mention this as well. This is where the transition takes place from stuck to unstuck and creates a closer bond with your audience, who can probably relate.

  End: This is the part where you are showing your audience how you can help them with their solutions. You provide the launch of your group coaching program by listing the features and benefits so they know what kind of results they will achieve when working with you.

- **Authenticity:** Ensure that when you describe your story from start to finish, you use real-life examples from when you started your business to the present. You can show behind-the-scenes images of how you started building the contents of your coaching program, client wins, and before-and-after photos.

- **Multimedia:** You will want to provide images, video testimonials from past clients, and digital mockups of the product launch.

This will provide more credibility for your business, and that the product is legit.

- **Practice:** You will need to create your webinar on a slide deck in Canva or in a Powerpoint platform. You would add bullet points of the main speaking points you will share with your audience. You will need to practice your speech and have notes to elaborate more on what you are planning to say in your entire presentation. Practice enough until you have memorized your presentation to avoid stalling since the webinar is pre-recorded or done live each time.

- **Engage and interact:** When presenting on a webinar, keep the engagement level consistent. You should ask questions that relate to the content to ensure your audience is engaged and not bored. You should also have a question-and-answer period at the end of the webinar for anyone wanting to learn more about your coaching program.

- **Analyze:** You can require your audience to fill out a survey or questionnaire form at the end of the webinar. This will help you see what feedback your audience provides and whether you need to make any changes or adjustments. You can also analyze the engagement to see which questions had more or fewer responses.

How to use storytelling in your social media

1. Understand your audience
2. Tailor content to your audience's needs
3. Provide value
4. Be authentic
5. Tell stories
6. Use variety in your content
7. Be consistent

CHAPTER 9

PR Marketing

What is PR marketing?

Public relations marketing combines the principles of public relations and marketing to create an effective strategy for promoting an organization, its products, or its services through different communication channels.

PR marketing consists of:

Media Relations: You scope out journalists, bloggers, and influencers to maintain and build relationships with them and create overall publicity.

Pitching to media outlets for more exposure: You would use your media kit or portfolio to collaborate with brands and pitch to companies to secure magazine spaces, newspaper articles, TV and radio guest appearances, and online media.

Content creation: You can create blog posts, press releases, magazine articles, podcast interviews, and solo or co-authored books. Before you create the content, ensure that your content aligns with your brand message and aligns as well with your target audience.

Influencer marketing: Influencer marketing has gained a lot of attention over the last few years. To level up your brand, you can collaborate with social media influencers with a high following to endorse your product or service. Their high following gives you the chance to amplify the brand message. Collaborating with other influencers benefits introverted business owners because they can leverage another audience within the same niche.

Social media engagement: You can create shareable content and community engagement on different platforms to keep your audience updated on upcoming projects and the latest news.

Event management: You can start attending virtual events to get more online visibility. There are such things as:

- Audio summits
- Video speaking summits
- Bundle events

Brand storytelling: Use your storytelling skills to showcase your brand's mission, values, and identity.

What are the benefits of PR marketing?

Increases brand awareness: Investing in your own PR marketing campaigns helps you increase brand awareness and gain more exposure and visibility.

For example, if you wanted to help introverts enter the world of digital marketing, you could advertise using PR marketing strategies, such as appearing on podcasts to discuss its benefits.

You can use well-known magazines to write articles about your business and why working with you enhances your business skills.

Cost-effectiveness: PR marketing is more cost-effective than using social media to run ads. It relies on media where there is already a high client base established.

If you were to get interviewed on a podcast, chances are the podcast already has a lot of subscribers and is repurposed on other marketing platforms. You can also pay a small fee to the podcast host to add a sponsored ad mid-recording so you can advertise your business or services.

If you rely solely on social media, you can use Facebook and Instagram to find leads and make sales. However, if you are a first-time business owner who wants to increase your visibility, do not invest in paid ads until you have already made consistent sales.

The reason is that paid ads can get expensive since you would need to spend money every day to run the ad. You may or may not get a lot of leads from your paid ads, depending on how many times a day and how much you pay to run ads.

Enhanced relationships: You will have the opportunity to build relationships with reputable brands to get more credibility presently and in the future. If you create effective long-term relationships, it is more than likely you will get possible referrals, which leads to extra publicity.

How PR marketing can benefit first-time introverted entrepreneurs

PR marketing can be very beneficial in providing strategies and tools so that entrepreneurs can build their brands without having to step too much out of their comfort zone. Here is a perfect example of how you can promote your launch using the power of PR marketing:

Example: You are getting ready to launch budget-friendly pre-made meal plans to help busy moms.

Content creation: You start a blog to write about the benefits of having budget-friendly meal plans as a busy mom. What kinds of meal plans will they have access to weekly, and how will this encourage healthy living in their home? You would repurpose the content onto social media platforms to get the word out.

Press releases: You would write a press release about the launch of your product and distribute it to different media outlets. Since your niche is health and lifestyle, you would write food blogs and articles in food and lifestyle magazines to launch your profitable product.

Influencer marketing: You collaborate with other influencers who have many mom followers who need time-saving tips.

Podcast interviews: You can pitch yourself as a guest on other people's podcasts to discuss your business. As an introvert, they can possibly feel more at ease when doing an audio interview in the comfort of their own home in a relaxed environment.

Video summits: To increase your visibility, you can participate in virtual video summits. If you are comfortable showing your face on camera, you can join an event where different speakers speak on a relatable subject.

Audio summits: An audio summit is similar to a podcast. It is a virtual event where you apply to become a speaker on a specific topic. For instance, I participated in an audio summit to discuss how my lead magnet can help their audience. That audience got a freebie to download and a chance to listen to the creator of the freebie to get more valuable insight. Audio summits are great for anyone who wants to listen to the content on the go.

Bundle virtual events: If you want more marketing and visibility, you can apply to contribute to a virtual bundle event. A bundle consists of business owners submitting their free or paid lead magnets to grow their audience. Usually, this would be to build their email list to funnel new subscribers. Bundle events are a great start for first-time introverted entrepreneurs who want to get in front of an audience.

Person and virtual networking: You can attend networking events, whether in person or virtually, to connect with other like-minded people. You should search online for upcoming virtual and in-person events. If you want in-person events, you can join the meetup groups in your local area to collaborate.

Magazines: You can apply to submit an article to a magazine within the same niche. For example, if you want to write an article on decluttering your closet in a tiny space, working with Life & Style Magazine would be best.

CHAPTER 10

Digital Products and Email Marketing

What are digital products?

Digital products are exactly what they are called. They are products created for digital use only. Digital products can benefit your online business, and here is why.

- **Digital products can create recurring revenue:** In the last few years, digital products have been taking over the digital marketing industry. With digital products on the rise, you can create your own products and sell them to your audience to make money. We will get into the different types of profitable digital products in just a moment.

- **Digital products are great resources for building an email list:** If you want to attract customers to your website, you can create an ebook or template to give away for free in exchange for their email address.

- **You can create digital products with master resell rights (MRR):** As a first-time entrepreneur, this is new to you, so it is important for you to fully understand the meaning of selling your product with MRR. For instance, if you create an ebook to sell to your audience, you can offer the master resell rights for them to recoup their profits at 100%.

- **Digital products can convert your leads into sales:** You can create custom digital products to convert those leads into sales. One way to do so is by creating a digital course or monthly membership program to help them with solutions to their problems. This is especially beneficial for introverts because they can feel more comfortable, hosting a small community of like-minded people.

What are the different types of digital products that can be created and sold?

There are many types of digital products that can be created by you to bring more visibility to your business. Here are the different types of digital products you can start using today.

- **Ebooks:** An ebook is an electronic guide that has bite-sized information to provide enough value on a particular topic.

  For example: You can create a 6–10 page ebook on a topic that appeals to your audience. If you are trying to start a blog for your business, you can find a reputable blogger who has free or paid ebooks to teach you all about starting a blogging business.

  If you are in a certain industry such as a virtual assistant looking for clients, you can create an informative ebook on *why you need a VA in your business!*

  You can list the top 3 or 4 reasons that a person will need your services. This will benefit you as a VA because you can provide value and add a call to action to either book a call with you or check out your website for more information.

- **Templates:** Templates are fillable sheets that you can use to provide to your audience as part of another coinciding digital product or sell individually. They can consist of journals, weekly, monthly, and daily planners, workbooks, and worksheets.

- **Mini-courses:** You can create mini-courses on a specific topic to sell to your audience. You can use your ebook to give minimal value and then use your mini-course as an upgrade to bring your audience to your online space. Mini-courses must be a condensed version of a full-blown course and focus only on a specific area. A good example would be a mini-course on how to make delicious sourdough bread.

You can divide the content into 4 modules, each breaking down the process step by step in each module, from what sourdough bread is to what you will need to start the process to the baking recipe to the final results.

- **Digital course:** You can create a fully digital course with more in-depth information. Your information could be a membership, a coaching cohort for a certain number of weeks, a self-paced study course, or a year-long mastermind. A digital course is made up of at least 8–12 modules, and you can either expand from the subject of your mini-course, or you can have multiple sub-topics to support the main topic. A digital course can be live or a pre-recorded training, and it can be held over a certain number of weeks or months.

- **Membership:** A membership is an ongoing platform where members pay monthly or annually to participate in a community to learn, engage, and share their expertise.

Digital products are a great asset to your business and the best way for your ideal clients to get a hold of your digital goods.

One of the most common strategies is to set up an automated sales funnel to turn those warm leads into hot leads. One way to convert your audience into hot leads is by implementing email marketing.

What is email marketing?

Email marketing is a marketing strategy that uses emails to engage and convert your potential leads into sales through a series of emails.

How does it work?

Before email marketing, you will have to create a digital product, which can be any of the above examples.

Once you decide which type of digital product you want to use to help solve your audience's problems, you will start putting the content together, remember, you are providing this free product to show your audience that you are the expert in their niche.

You will want to create an opt-in form to collect their email address and name in exchange for the digital product, also called a lead magnet. Once they provide the information required for the lead magnet, your email address will get set up on an email platform. The host of the lead magnet will have set up an email sequence to welcome you into their world and take you through their business journey.

The welcome sequence would usually be a total of 6–8 emails that are automated and take you through a process of developing a working relationship with you.

Email 1#: This is where you thank your audience for signing up for your lead magnet and make sure they receive the link. You will also provide a valuable tip from the lead magnet and include your social links at the end to stay connected with you.

Email 2#: The second email is where you will continue to provide more value based on the lead magnet. This is where you can speak about your personal struggles with how you got started in your business, the challenges you faced, and how you got the results.

Email 3#: You can introduce yourself and write about who you are, who you help, and how you got started in your business. Your experiences when you started then and now.

Email 4#: The myths and facts in your industry or what are the biggest fears the industry is going through.

Email 5#: Introduce your medium or high-paying offer. If you are a coach or a service provider, you want to start talking about your latest offer that can help them get from stuck to unstuck. You want to include bonuses and why they need to work with you.

Email 6#: Questions and answers about the offer. You can encourage your audience to reply to your email with questions they may have. You can also do a question and answer email where you are answering the most common questions people are asking.

Email 7#: Friendly reminder that the offer expires in 48 hours! You want to remind the subscribers they have limited time to register.

Email 8#: Last day before the offer expires. This gives a sense of urgency for them to hurry up and purchase the offer before it's gone.

If by the 8th email, you still have not gotten any leads to your offer, you can continue to nurture them by writing a follow-up email to ask why they didn't register for the offer. This can help provide feedback to know if the offer spoke to them or not.

You will have to keep in mind that your automated welcome sequence will not always have your audience warmed up right away and ready to buy. If they chose not to buy your offer initially, it is most likely because you need to continue nurturing them first. You can switch them over to your regular newsletter sequence, where you provide valuable tips and keep them up to date with low- and medium-ticket offers.

Introverts can benefit from all of these different strategies if they want to succeed in the online space. They can find more adaptable ways to get more creative when it comes to showing up online.

About the Author

 Julie Lavia is a visibility strategist and digital marketer dedicated to helping first-time female entrepreneurs launch and grow their online businesses with confidence. As an introvert herself, she deeply understands the unique challenges introverted women face when stepping into the digital space and is passionate about guiding them toward success.

A strong advocate for female empowerment, Julie encourages women to share their stories, embrace their voices, and build a brand that truly represents them. Based in Montreal, Canada, she is a self-proclaimed foodie who loves exploring diverse culinary experiences whenever she gets the chance.

Facebook: www.facebook.com-/julie.lavia.1/

Instagram: https://www.instagram.com/julielaviacoaching

Website: www.juliechristinacreates.com